Metro Marvels

How Cities are Embracing Green Solutions

Table of Contents

Chapter 1. Introduction

Step into the future with us as we explore the effervescent revolution taking place in concrete jungles around the world in our Special Report: "Metro Marvels: How Cities are Embracing Green Solutions". With a tone that is as upbeat as the topic itself, we will guide you through the inspiring transformations our urban landscapes are undergoing in their quest for sustainable living. This comprehensive report offers an enthralling insight into breakthrough initiatives, innovative technologies, and the indomitable spirit of city-dwellers charting a greener path for all of us. It's not just a story of the cities, but a story of humanity taking giant strides towards a sustainable future. This isn't just a read, but your ticket onto a thrilling roller coaster of invigorating changes, potential solutions, and forward-thinking urban development. Prep yourself to be amazed and uplifted, and dare we say, ready to invest in this exclusive Special Report today!

Chapter 2. Skyward Solutions: Vertical Greenery in Urban Design

Concrete metropolises around the world have started to re-define their relationship with nature and it's easily evident in the skyrocketing trend of Vertical Greenery. As urban areas burst at the seams, designers and architects are looking upwards and finding unique ways to integrate greenery into city buildings with startling results.

2.1. The Ascent of Vertical Greenery

Vertical Greenery, also known as vertical gardens or living walls, is a phenomenon that is as innovative as it is rooted in ancient history. The concept of gardens that grow vertically rather than horizontally has its genesis in the ancient civilizations of Babylon, seen in the mythical Hanging Gardens. Fast forward a few millennia, and this greening strategy has been embraced by modern urban design, transforming gray cityscapes into verdant vertical forests.

A vertical garden involves the incorporation of flora - trees, shrubs, flowers, and a variety of other plants - on, and in the façade of buildings, retaining walls or stand-alone structures. The benefits are numerous, offering a plethora of environmental and human health advantages, thereby enhancing the livability of our cities.

2.2. The Green Benefits of Vertical Gardening

The benefits of vertical gardening are vast, offering a multitude of environmental and socio-economic advantages. By literally building flora into and onto structures, cities can enjoy cleaner air, reduced temperatures, and amplified biodiversity.

Air purification is a critical feature of vertical gardens. The urban jungle contributes significantly to air pollution. Constructing vertical gardens assists in filtering particulate matter out of the air, absorbing toxins and, consequently, improving air quality.

Vertical gardens reduce urban temperatures by improving a city's microclimate through the natural process of evapotranspiration. It is estimated that one square meter of vertical garden can absorb the same amount of solar radiation as five square meters of traditional roof gardens.

Moreover, this integration of vertical greenery invites biodiversity. It provides a habitat for insects, birds, and other small species, fostering a more diverse ecosystem right in the heart of a bustling cityscape.

2.3. A Pillar of Green Design: Architectural Integration

"Farming the facades" is becoming increasingly popular as urban architects find creative ways to integrate greenery into building design. It can involve something as simple as climbing plants on walls, or something as complex as planned green facades with dedicated irrigation and drainage systems.

Renowned Italian architect Stefano Boeri is known for his ambitious vertical forest projects, drastically transforming concrete facades into living, breathing entities. His project, Bosco Verticale (Vertical Forest) in Milan, incorporates more than 20,000 trees and plants on its facades.

In Singapore, the PARKROYAL on Pickering is another sterling example of this practice. Designed by WOHA Architects, the hotel incorporates sky gardens, reflecting pools, waterfalls, and planter terraces, creating a hotel that is as much a garden as it is a place for people.

2.4. Overcoming the Challenges

While the concept of vertical greenery is compelling, its application isn't without obstacles. Constructing these living facades requires complex design considerations including weight, irrigation, maintenance, wind load, and plant selection.

The high initial costs, ongoing maintenance expenditure, and possible structural adjustments form part of the barriers that architects and property developers need to manage. Ongoing research and successful examples like the Bosco Verticale demonstrate the feasibility and long-term benefits of investing in such green solutions.

2.5. The Future is Green... and Vertical

The vertical greening revolution is underway, and as architects and urban planners become more adept at integrating living walls into their designs, our cities will become both more beautiful and healthier places to live. Skyward solutions like vertical greenery provide one more powerful tool in our environmental toolkit, leading us toward a brighter, greener urban future.

Urban design will continue to reach new heights, quite literally, as it incorporates vertical greenery. As we ascend higher and stand amidst these thriving green skyscrapers, we realize that the future of urban sustainability is not just about preserving the green we have, but cultivating more wherever we can. And where better to start, than our very own 'concrete jungles'.

Just imagine looking out of the window of your high-rise apartment, not overlooking a cold, gray cityscape, but a lush, verdant vista. This is the promise of vertical greenery. This upward trend suggests that the cities of the future will not be monolithic steel mountains and endless pavement, but green, living organisms where nature and urban design exist in harmony.

Chapter 3. Roads of the Future: Green Streets and Eco-Transport

As the sun rises on a city of the future, the vision is dazzling. Buildings that touch the sky are covered in a sheen of solar panels, parks sprout from rooftops, and the streets below bustle with electric vehicles (EVs) and cyclists. At the heart of this urban sphere, you find the revolutionary concept of green streets and eco-transport, which sets the stage for a sustainable future.

3.1. Developing Eco-Streets

Every journey starts with a single step, or in the case of cities, a single road. Green streets are rooted in an approach that integrates stormwater management into street design. This progressive idea combines engineering, urban design, landscape architecture, and ecology. It goes beyond the usual gray infrastructure to include spaces and elements for rain and stormwater.

Cities like Portland in the USA have embraced eco-streets, capitalizing on multiple benefits such as improved water quality, more vibrant streetscapes, and resilient urban ecosystems. The novel incorporation of plants and porous pavers into the design helps in managing stormwater runoff, supports wildlife, and promotes active transportation.

key aspects of eco-streets

- Stormwater management

- Vibrant streetscapes

- Promoting urban ecosystems

- Active transportation

Moving beyond the grass-roots level, the bigger picture of eco-streets is found in their continuous evolution, which thrives on fostering cycling and pedestrian-friendly spaces. From newly paved bike lanes to footpaths enhanced with trees for shade and beauty, cities are prioritizing people and the planet over vehicles.

3.2. Embracing the Cycle

Cycling has emerged as a symbol of green mobility and cities are reinventing the wheel with transformative measures. Amsterdam, renowned as the 'City of Bikes', boasts an extensive network of bike paths. Other countries like Denmark and Colombia have followed suit, encouraging cycling with dedicated infrastructure, thereby reducing carbon footprints, mitigating city noise, and improving citizen health.

3.3. EVolution of Transportation

Another groundbreaking development is the surge in popularity and deployment of electric vehicles. Whether it's an electric bike, a scooter, a car, or a bus, EVs are driving the shift towards cleaner, greener transportation. In 2021, global electric car stock surpassed 10 million, a major milestone in the fight against air pollution and climate change. Cities like Shenzhen have committed to fully electrified urban transit, replacing their entire fleet of 16,000 buses with electric ones.

3.4. Complete Streets

However, it's not just about eco-streets or eco-vehicles but marrying these two burgeoning concepts to create what enthusiasts call "complete streets." Complete streets incorporate eco-street design and transit-oriented development. They are engineered to prioritize pedestrian and bicycle traffic, transit accessibility, and greenery over automobile traffic. They embody a holistic vision of future roads, an amalgamation of greenery, active travel, and public transportation, creating more shared and safe spaces for all road users.

3.5. Implementing Green Solutions

Now, as we envision cities of the future, how do we get there? Sculpting this sustainable urban environment requires strategic planning mixed with innovative solutions. From using recycled materials for road construction to integrating tech-based AI solutions for efficient traffic management, cities are being put under the revolutionary chisel.

Invisible solar panels on the road surface, roads lit by phosphorescent materials, and wireless charging lanes for electric vehicles are some of the ingenious inventions. Additionally, green public transit systems, green freight initiatives, and wide-scale bike-sharing systems are being adopted to achieve sustainable urban mobility.

3.6. Future Roadmap

As we embark on these green streets, it's clear that the road to the future lies in incorporating environmental considerations in city planning, overcoming resistance to change, fostering innovation, and nurturing an attitude of respect towards nature.

While it's a monumental task, cities worldwide have started adopting these strategies. They are learning from each other and adapting to their geographical and social contexts. Moreover, with increased citizen participation and collaboration between local governments, private entities, and environmental organizations, the transformation to green urban landscapes can be realized.

The green streets and eco-transport of the future is a testament to the indomitable spirit of humanity. As we persistently strive to combat climate change and create sustainable habitats, it is evident that the language of green solutions is global and universal. The vision of eco-roads, bursting with possibilities and innovations, guarantees one thing - the journey towards a sustainable future is as exciting as the destination.

For an insightful journey into the revolutionary developments in the sustainable transport space, make sure you stay tuned for our next chapter, which takes a deep dive into public transportation's evolution and the strides towards creating more environmentally conscious mobility solutions. Through our comprehensive special report, we aim to equip you with all the knowledge and tools needed to partake in this grand journey towards sustainability.

The green roads await your arrival; let's begin this exciting ride towards a future where cities dance to the rhythm of sustainability. There couldn't be a better time to be a part of this exhilarating revolution.

Chapter 4. The Ripple Effect of Rooftop Gardens

The metropolises of the world, known for their soaring skyscrapers and sprawling urban landscapes, are increasingly seen sporting a coat of green on their rooftops. This pictorial change is not merely a decorative whim, but a purposeful shift towards a greener, healthier and more sustainable future. In this chapter, let's take a look at how cities across the globe are experiencing the ripple effect of rooftop gardens.

4.1. The Development and Purpose

Rooftop gardens, also known as green roofs, green tops, or sky gardens, are essentially vegetated surfaces on top of buildings, big and small. They've been around for centuries, but have witnessed a tremendous spike in popularity over the last few decades. Their purpose goes beyond prettifying city skylines; they are a practical response to growing environmental concerns and urban challenges.

In bustling cities where space is a hot commodity, rooftops provide an under-used expanse perfect for reclamation. Rather than hosting satellite dishes or billboards, these spaces can support lush gardens that bring about a host of benefits. They help in improving urban biodiversity, enhancing air quality, managing stormwater runoff, insulating buildings, and even promoting community cohesion and mental wellbeing.

4.2. The Environmental Impacts

One of the primary benefits of rooftop gardens is their role in countering the Urban Heat Island (UHI) effect. UHI is a phenomenon wherein built-up areas are significantly warmer than their rural surroundings. Green roofs tackle this issue by providing shading and evapotranspiration, thereby lowering overall city temperatures and reducing the necessity for air conditioning, which in turn decreases greenhouse gas emissions.

Furthermore, new research indicates that rooftop gardens can play an instrumental role in sequestering carbon dioxide, a key contributor to global warming. The vegetation on green roofs absorbs CO_2 during photosynthesis, storing carbon in the plants themselves and the growing medium.

Rooftop gardens also play a key role in stormwater management. In dense urban areas with expansive concrete streets and buildings, rainwater has nowhere to infiltrate, leading to increased runoff and potential flooding. Green roofs can capture and hold a significant portion of rainwater, preventing its instantaneous run-off and minimizing stress on city sewer systems.

4.3. Social and Economic Benefits

Rooftop gardens aren't just good for the planet; they're also beneficial for the people who live and work around them. Green spaces are now considered a scarce commodity in many crowded cities, which places a premium on quiet, serene spaces enveloped by nature. Rooftop gardens can serve as communal spaces, providing residents with a place for relaxation, socializing, and even food production.

Several cities have initiated urban agriculture projects on rooftops that result in locally grown, pesticide-free produce. In combination with the potential for job creation and improving food security, these initiatives often inspire a sense of community ownership and pride.

Economically, green roofs offer advantages too. Although the upfront costs of installing rooftop gardens can be high, the long-term benefits make it a sound investment. They can reduce the need for HVAC systems, leading to substantial energy savings. Further, the protective layer of a rooftop garden can increase the lifecycle of the roof membrane, reducing maintenance and replacement costs.

4.4. Want to Harness the Ripple Effect? Here's How

The question remains: How can a city or a building owner successfully implement a rooftop garden project? From understanding the structural strength of a building to choosing the right plant species, and navigating building codes and permits, it's a process that requires thoughtful planning.

Teamwork and collaboration among architects, environmentalists, policymakers, and local communities play a crucial role. Innovative technologies such as green roofing systems and hydroponic farming methods offer promising solutions. Equally important is public engagement and awareness, as well as the integration of green roof policies into urban planning and design guidelines.

In countries like Germany and cities like Toronto, strong public policies and incentives have led to a high proliferation of green rooftops. It serves as an excellent example for urban centers worldwide aiming to transform their grey horizons into green canopies.

4.5. Conclusion: The Seeds of a Revolution

The humble rooftop garden, often overlooked in urban planning, has emerged as a potent tool in city dwellers' hands to reclaim their environment. This green revolution, spreading its roots across global cities, sends a powerful message: sustainable living is not limited to countryside locales or dedicated eco-villages. It can, and indeed should, flourish in the heartland of human civilization: our cities.

As cities continue to grow and the world's population becomes increasingly urbanized, the small, unassuming rooftop garden will continue to play a critical role. The ripple effect of these elevated green spaces may just serve as the tidal wave that leads our urban areas towards more sustainable and resilient futures. Indeed, cities worldwide are not just witnessing a revolution; they are actively participating in it - one rooftop garden at a time.

Chapter 5. Inside the Green Machine: Sustainable Infrastructure Innovations

The skyscrapers shooting upwards, their reflections gleaming on glass surfaces, are no longer just icons of human ambition; now they have another story to tell, one of sustainability and green living. Noted as urban spires of resilience, these modern edifices are adopting eco-friendly designs paving the way for the next phase of urban development, one that's characterized by ecological responsibility as much as by architectural creativity.

5.1. Building With Nature: Biophilic Design

In our quest for sustainability, we have discovered an ingenious solution in the unlikeliest of places: nature herself. Biophilic design, also known as 'building with nature,' involves integrating natural elements into architectural design. This can range from the creation of green roofs and walls, installation of skylights for natural light, usage of nature-inspired materials and textures, to the inclusion of open spaces filled with vegetation within the building structure. These not only reduce energy consumption by providing natural light, insulation, and temperature regulation but also contribute to improved mental well-being of the inhabitants. The reduction in energy consumption tangibly translates into a significant cut on carbon emissions, aiding in our combat against climate change.

5.2. An Edifice of Energy Efficiency: Green Buildings

Any discussion about sustainable infrastructure would remain incomplete without mention of green buildings. Green buildings are designed and constructed to minimize environmental impact and maximize sustainability. They accomplish this through energy efficiency, water-saving measures, waste reduction, use of recycled and renewable materials, and provisions for natural light and ventilation. Green building certifications like LEED (Leadership in Energy and Environmental Design) have set benchmarks for implementing and recognizing green practices in building design and operation. The increasingly widespread popularity of such certifications confirms the steadily growing commitment to sustainable living across geographies.

5.3. The Future of Highways: Green Roads

Roads often eat up huge expanses of land and massively disrupt ecosystems, but they too are part of the sustainable infrastructure revolution. Roads might soon become green energy generators, thanks to innovative concepts like solar highways and piezoelectric roads. Solar highways, constructed using photovoltaic pavement, make productive use of the space by harnessing solar energy. This captured energy could fuel roadside lamps, traffic systems, and even electric vehicles. Heavier traffic on piezoelectric roads generates more electricity as piezoelectric crystals embedded in the road convert mechanical stress into electrical energy.

5.4. Blue Green Infrastructure: Melding Urban Development and Ecosystem Services

Of particular interest is the concept of Blue Green Infrastructure (BGI), a network that integrates water management, biodiversity conservation, and urban green space provision into urban development. This concept helps manage flood risks, improves air and water quality, promotes biodiversity, and offers recreational benefits too. It signifies a broader, more inclusive perspective towards urban sustainability where green spaces are not just aesthetically pleasing but multifunctional, rendering a multitude of services including stormwater management and urban heat reduction.

5.5. Technology and Data: Smart Sustainable Cities

As digital technology increasingly permeates all facets of human life, it's no surprise to find it at the core of sustainable city development. Smart cities leverage technology and data to optimize resources, enhance governance, and improve the overall quality of urban life. The integration of IoT (Internet of Things), AI (Artificial Intelligence), 5G, cloud computing, and big data into urban planning yields smart grids, intelligent transportation systems, intelligent waste management, and smart buildings. The result is a comprehensive sustainability model that runs on real-time data and responsive planning, paving the way for resilient, efficient, and inclusive urban landscapes.

As humans living in an increasingly urban world, these sustainable infrastructure innovations offer not just reasons for hope, but tangible, groundbreaking solutions. They herald an era in which skyscrapers are more than just vertical cities; now, they are also trees, and roads do more than connect places; they generate energy. In this ecological urban renewal, cities are becoming prime sites for sustainable development, with human creativity and technology joining hands with nature to pioneer a greener future.

Chapter 6. Energy Reborn: Renewable Power in the Metropolis

For too long, cities across the globe have been infamous for their towering skylines, glinting with non-renewable energy consumption. Today, though, those very metropolises stand at the forefront of an energy revolution. They are embracing the power of renewable energy, emerging as vital engines of clean power generation. Let's delve into their journey—understanding their transformation, the technologies powering this seismic shift, and the consequences of their brave actions.

6.1. The Rising Need for Green Energy

In the face of environmental peril and an increasing global temperature, the call to combat climate change is becoming more urgent than ever before. It's not an option, but a necessity—a seismic shift from non-renewable to renewable energy sources. The International Energy Agency (IEA) has recognized this growing need, suggesting that to limit global warming to 1.5 degrees Celsius—consistent with a sustainable energy pathway—global renewable energy capacity will need to expand significantly by 2050.

Cities, being major consumers of energy, have risen to the challenge, boldly paving the way for this green transformation by employing renewable energy sources: sun, wind, water, and more.

6.2. Embracing Solar Power

Perhaps the most widely recognized form of renewable energy, solar power, has seen a surging deployment in cities worldwide. Even the most gnarly concrete jungles are now bedecked with gleaming solar panels, transforming rooftops, vacant lands, and degraded brownfields into valuable energy assets.

Modern cities like San Francisco have taken the lead with legislations requiring new buildings to have solar panels installed, thereby invigorating the green energy movement. From generating power locally to creating jobs and reducing reliance on fossil fuels, the sun-drenched path is ushering an era of widespread energy and economic benefits.

At the heart of this solar revolution is technological innovation that increases efficiencies and reduces costs. Advances like bifacial solar modules, which produce electricity from both sides, and solar tracking systems contribute to harnessing the sun's power more effectively.

6.3. Harnessing the Wind

As the cost of wind energy has dropped, city-dwellers have begun harnessing its power with great enthusiasm. Both onshore and offshore wind farms are seeing rapid expansion in urban areas worldwide. These establishments not only signify an earnest adoption of green solutions but also enable cities to become less reliant on external supplies for their electricity needs.

Offshore wind farms, despite being more expensive and logistically challenging to build, offer colossal energy potential. Cities like Copenhagen have capitalized on their coastal locations, installing large wind turbines in waters at a significant distance away from urban areas to assuage noise concerns while supplying green energy to their inhabitants.

6.4. Hydropower's Urban Promise

Hydropower, one of the oldest known sources of renewable energy, is finding its place in the urban scene. Cities that boast of rivers and tides are employing small, turbine-based systems to extract energy from flowing water. Some use this green power to light public areas, while others sell it back to the national grid.

Bristol, a port city in England, is a paragon of such initiatives, reclaiming the renewable pathways by pilot-testing water wheel technology for power generation.

6.5. Biogas from City Waste

Cities produce an enormous amount of waste, giving rise to significant environmental challenges. Innovative city planners are turning this bane into a boon—extracting biogas from organic waste.

Biogas plants within city parameters convert organic waste from households, gardens, parks, and commercial sources into usable methane-rich gas and fertilizers. Stockholm, with some 6,000 biogas-powered vehicles, shows how cities can use waste to their green advantage.

6.6. Energy Storage: The Balm for Intermittency

Renewable energy's intermittency—its dependence on elements like the sun and wind—poses challenges to its continuous use. However, advances in energy storage solutions, like lithium-ion batteries, are alleviating these concerns.

Neighborhood-scale energy storage systems and energy-sharing platforms are changing the landscape of urban renewable energy usage, establishing a resilient, decentralized, and democratized power infrastructure.

6.7. The Impact of Renewable Energy Adoption

The shift to renewable energy sources offers a slew of benefits to our urban landscapes: from reduced greenhouse gas emissions and enhanced public health due to cleaner air, to economic benefits from jobs in the thriving green energy sector.

Moreover, as cities continue to grow, they can meet the increasing energy demand by harnessing and storing renewable energy—forging a path toward energy independence. This energy revolution is not a pipe dream, but a reality unfolding in our urban environments worldwide.

In conclusion, metropolitan areas worldwide are standing at the forefront of an energy revolution, ensuring a sustainable and resilient tomorrow. Their innovative efforts and positive strides are not only transforming the way we view urban landscapes but blurring the lines previously dividing economic development and environmental preservation. It's an incredible journey from being energy consumers to becoming energy generators—a journey we can all take inspiration from as we prepare for a greener future.

Chapter 7. Waste Not, Want Not: Effective Urban Recycling Systems

In the shifting tides of urban sustainability, proficient recycling systems have severed their ties from being a trend, evolving instead into an incontrovertible necessity. Contemporary metropolitan areas worldwide are tapping into innovative solutions to tackle the gigantic problem of waste management, bending the arc of development towards a sustainable and healthier future.

7.1. Waste Management: A Pivotal Urban Challenge

Submerged in the waves of technological revolution and modernization, urban areas are fast becoming the pulse of human civilization. As we applaud the strides of progress, an unattended byproduct - waste - poses severe challenges.

Most urban dwellers barely give a second thought to their trash once it's in the dumpster. The reality, however, is far more complicated. As per the World Bank, urban areas generate 2.01 billion tonnes of solid waste annually, projecting a daunting increase to 3.40 billion tonnes by 2050.

7.2. Impacts of Inefficient Waste Handling

If not managed proficiently, the heaps of urban waste could dwarf the towering achievements of our civilization. It could lead to grave issues like soil degradation, water pollution, and air contamination, imposing colossal health and environmental hazards.

The World Health Organization (WHO) estimates that improper waste management contributes to about 5% of all global deaths. Widespread issues such as respiratory problems, gastrointestinal disorders, and various types of infections afflict people living in areas with unmanaged waste.

7.3. Embracing Modern Methods: The Paradigm Shift

The staggeringly morose situation calls for a turnaround. Plunging headfirst into innovative endeavours, modern cities are now striving to transform urban waste from a concerning liability into a viable resource. The battle is no longer restricted to 'managing' waste; the mission now is to 'recycle,' to extract value out of something otherwise discarded.

7.4. Innovative Approaches to Recycling

Several cities worldwide are adopting pioneering initiatives for effective waste recycling, pushing the boundaries of what's typically perceived as waste management. From tech-led solutions to community-driven programs, each initiative translates to a step closer towards sustainable urban living.

Sweden, for instance, has executed an efficient waste-to-energy program that converts household waste into energy for heating and electricity. In San Francisco, the Zero Waste Program sets a goal to send nothing to landfill or incineration by 2030.

More than a top-down approach, effective recycling systems also require grass-root participation. The Bangalore-based initiative

'2b1b' (2 bins, 1 bag) has been driving civic participation in waste segregation, significantly boosting recycling efforts.

7.5. Technological Innovation in Urban Recycling Systems

Technology plays an instrumental role in upgrading waste recycling systems. Advanced waste sorting technology, AI-powered waste segregation tools, and data analytics for managing waste collection form the backbone of modern recycling efforts.

For instance, AI-enabled smart bins in Dubai automatically sort and compact trash, maximizing recycling efficiencies. In Copenhagen, sensor-enabled garbage cans send notifications for pick-ups when they are nearly full, optimizing waste collection routes.

7.6. The Future of Urban Recycling Systems

While we have come a long way in overhauling urban recycling systems, the path ahead is still steep. Education and awareness about waste segregation, incentivizing recycling, stringent waste disposal laws, and further tech integration are some facets warranting attention.

As cities continue to expand, urban recycling systems hold the key to sustainable living. The shift from a vicious cycle of waste generation and disposal to a virtuous cycle of waste creation, segregation, and recycling is gaining momentum. In this inspiring transformation, technology and community engagement are turning out to be the real game-changers. And as various city-led initiatives demonstrate, urban recycling is no longer a realm of mere possibility; it is quickly becoming an outstanding reality.

To ensure future generations inherit a world rich in opportunities, not overwhelmed by waste, we need to champion these advancements in recycling systems. They break the chain of neglect, bringing in a fresh perspective of respect for every scrap we throw away and reaffirming our faith in the power of innovation to resolve seemingly insurmountable problems. The revolution is here, in our hands, waiting to be nurtured for a cleaner, greener, and more sustainable future.

Chapter 8. Blue Skies Ahead: Air Quality Improvement Strategies

The crucial importance of air quality goes without saying. Every breath drawn contributes to our vitality, powering everything from our cells to our cognitive functions. To illustrate this, its counterpart, air pollution, is a grave global concern, with the World Health Organization (WHO) stating that it causes an estimated 7 million premature deaths annually. The positive side of this grim statistic? There's a lot of room for improvement and, excitingly, cities around the world are making dynamic strides towards clearer, cleaner skies.

8.1. The Air We Breathe: Air Quality and Health

Air quality impacts many aspects of our health. From respiratory conditions like asthma or bronchitis, to cardiovascular diseases, the quality of the air has direct implications on our well-being. Pollutants such as sulfur dioxide, nitrogen dioxide, and particulate matter are particularly detrimental, causing a plethora of health issues. Beyond this individual toll, poor air quality strains healthcare systems and economies through increased medical costs and reduced workforce productivity.

In light of this, many cities have started to refine their strategies for air quality improvement. From embracing green spaces, to enhancing public transport, and supporting renewable energies, these urban environments are taking a proactive stance.

8.2. Urban Forests: A Lush Lunge for Healthier Air

Cities like Melbourne, Australia and Milan, Italy are investing in an innovative, yet beautifully simple strategy: urban forests. These cities are aiming to plant millions of trees within their boundaries. Trees absorb carbon dioxide, filter particulate matter from the air, and release oxygen, making them nature's own air purifiers. Melbourne aims to double its canopy cover to 40% by 2040, while Milan hopes to have 3 million new trees by 2030.

This isn't just a beneficial move for air quality. Urban forests produce cooling effects, mitigate urban heat island effect, increase biodiversity, and improve residents' mental health. With all these advantages, it's no surprise that city planners are eager to green their concrete streets.

8.3. The Power of Public Transport

Public transportation holds tremendous untapped potential for improving urban air quality. Cities like Copenhagen, Denmark and Curitiba, Brazil have robust, efficient public transportation systems that significantly reduce the use of personal vehicles, hence reducing the volume of pollutants.

Germany's Stuttgart provides an exceptional example in this respect. The city operates an integrated mobility platform that combines public transport, car-sharing, bike rentals, and pedestrian routes in one seamless system. This encourages residents to opt for cleaner modes of transport, contributing positively to the city's overall air quality.

8.4. The Clean Energy Transition

Switching to renewable energy sources is pivotal in the fight against air pollution. Solar, wind, and hydro energies do not produce harmful emissions and are becoming increasingly cost-competitive.

Cities like Reykjavik, Iceland and Masdar, UAE, are leading the way using geothermal and solar power, respectively, as their primary energy sources. Reykjavik is especially notable, using geothermal energy for heating and electricity, practically eliminating energy-related air pollution. Similarly, Masdar aims to be a carbon-neutral city, relying solely on renewable energy.

8.5. Microbial Magic: Harnessing Microorganisms

A novel approach to improving air quality is reaping the benefits of microorganisms. Several microbes, like some strains of Pseudomonas, have the ability to metabolize pollutants, effectively 'eating' the compounds that taint our air.

Mexico City, named one of the most polluted cities in the world, has seen the construction of a 'smog-eating' building coated with titanium dioxide. This material, when activated by sunlight, interacts with pollutants and neutralizes them, cleaning the air one building at a time.

8.6. The Rise of Smart Cities

Technology is reshaping our cities in the form of the Smart City revolution. With advanced data collection and interpretation, cities are better equipped to detect sources of pollution and take effective countermeasures.

Barcelona, Spain is one such city, using its integrated sensor system to monitor pollution levels and traffic flow. Solutions like dynamically controlled traffic lights optimize vehicle flow, reducing idling time and resultant emissions.

From increased urban forestry to innovative use of microorganisms, the focus on enhancing our air quality is gaining momentum. Initiatives once thought of as futuristic are becoming concrete reality. Herein lies a hopeful message for us all: our cities are evolving, and cleaner, healthier air is on the horizon. Proactive, innovative, and committed, these urban areas exemplify humanity's capacity not only to adapt but to thrive in harmony with the environment.

Chapter 9. Urban Oasis: The Importance of Green Spaces

Escaping to the haven of a verdant park or a tranquil garden in the heart of a bustling metropolis is a relief many of us cherish. Amid the swarm of skyscrapers and an orchestra of honks, these green spaces are precious lungs of the city - sanctuaries that not only refresh the human spirit but also play a critical role in sustaining ecological health and reducing pollution.

9.1. Embodying Ecological Wellness

Urban parks, gardens, and playgrounds are an essential aspect of city planning, extending far beyond aesthetics and recreational purposes. These green spaces become focal pillars of the urban ecosystem, maintaining biodiversity and offering refuge for native flora and fauna. Birds sing from the treetops, squirrels scamper along branches, insects hum in the undergrowth, and a variety of plants create a lush, green palimpsest that embodies ecological wellness right in our concrete-dominated neighborhoods. As we further delve into their importance, it becomes clear that these lush expanses are indispensable pieces in the urban sustainability puzzle.

9.2. The Breathing Nodes of the City

Concrete and asphalt sprawl trap heat, causing urban heat island effect, where city temperatures can be several degrees hotter than the surrounding rural areas. Enter urban green spaces - the city's natural air conditioning unit. Trees and green zones alleviate this effect by reflecting sunlight and providing shade. Parks act as the city's lungs, consuming carbon dioxide, exhaling oxygen, and trapping harmful pollutants, making them crucial air-purifiers in an age of escalating pollution.

Moreover, these spaces are groundwater recharge points, sustaining the water table in an era increasingly haunted by the specter of water scarcity. Rainwater permeates the soil in these areas, reducing surface run-off, replenishing groundwater, and mitigating flood risks. These spaces truly encapsulate the symbiosis between humans and nature.

9.3. Enhancing Urban Life and Wellbeing

Green spaces contribute to our cities' social fabric as well, becoming epicenters for community engagement and physical activity. Active park users are more likely to achieve the recommended physical activity level than non-park users. In offering open spaces for sports, exercise, or simply a leisurely stroll, they bring people together in shared experiences, fostering community bonds and nourishing mental health.

Indeed, the benefits of green spaces to our psychological wellbeing are paramount. Studies report lower levels of stress, depression, and anxiety in residents with access to green spaces. The calm nature provides, it seems, is a balm in our age of hyper-connectivity and relentless pace that often drains vitality.

9.4. The Role of Urban Agriculture

In recent years, there's been a burgeoning trend of urban agriculture within city green spaces. Community gardens flourish across rooftops, disused lots, and parkland, producing fresh and local food. Urban agriculture fosters resilience, food security, and reduces the environmental footprint associated with long-distance food transportation. It's another testament to the multifaceted roles green spaces play in crafting sustainable cities.

9.5. Challenges and Future Directions

However, the provision and maintenance of green spaces are not without challenges. Urbanization, encroachment, and lack of equitable distribution continue to be persistent problems. Opportunities lie in developing innovative urban design solutions that marry urban densification with the creation and preservation of green spaces. Rehabilitating disused urban spaces into green zones could be a thoughtful way forward.

In conclusion, green spaces are the beating heart of our cities – vital to ecological health, social cohesion, physical well-being, and mental wellness. In our journey towards sustainable urban development, they remain invaluable. As we continue to innovate and experiment, ensuring that every city-dweller has access to these urban oases should be central to our city planning ethos, guiding us towards a greener, healthier future for all.

Chapter 10. Resilient Cities: Climate Change Adaptation Measures

As we stand on the precipice of climate change, poised to either rise to the challenge or succumb to its devastating effects, the metropolises around the world have begun to design and implement city resilience strategies that not only serve to safeguard their infrastructure and citizens, but also to cherish and enhance their unique urban ecosystems. Embracing a new age of sustainability and resilience, cities are boldly innovating, learning from each other, and adapting in order to prepare for the uncertain and potentially tumultuous times ahead.

10.1. Adapting to Rising Sea Levels

Cities located by the coasts are most vulnerable to climate change, as they often face the dual threat of rising sea levels and increased storm surge. To tackle these challenges, urban resilience strategies are incorporating both hard and soft infrastructure solutions.

The Netherlands, a country with much of its land below sea level, leads the world in water management and coastal defenses. Its Room for The River project, instead of resisting the water, has redesigned the landscape to live with the water, through river widening, floodplain restoration, and retention areas.

Another notable innovation in this arena is the use of amphibious architecture, a technique demonstrated in cities like Rotterdam and Ijburg in Amsterdam, which incorporates floating houses and buildings capable of rising and falling with the water levels.

10.2. Climate-proof Infrastructure

For cities, the key to climate change adaptation often lies in strengthening the resilience of their infrastructure. From dual-purpose open spaces that serve as flood plains during rains to permeable pavements that help manage stormwater, cities are leveraging innovative design principles to climate-proof their infrastructure.

Chicago's "cool roofs" initiative is a commendable example. To combat the urban heat island effect, the city started using reflective roofing materials that respond to fluctuating temperatures by changing colors, thereby reducing cooling costs and greenhouse gas emissions.

10.3. Urban Greening

Urban greening initiatives have become an integral part of any resilient city strategy, with trees and plants playing double-duty as air filters and natural water management systems, while also serving to reduce the urban heat island effect.

Singapore, the 'City in a Garden', has demonstrated how urban greening can be an effective climate adaptation measure through its extensive Greening Master Plan. In addition to laying out strategies for street and park trees, the plan also underscores the importance of vertical greenery and extensive roof gardens in high-rise constructions.

10.4. Taking Advantage of Technological Solutions

The advancement of technology has opened a plethora of opportunities for cities to enhance their resiliency. From early-warning systems for extreme weather events to 'smart grids' that allow cities to manage their energy demand and supply dynamically, high-tech solutions are playing an increasingly vital role.

New York City's PlaNYC initiative, for example, used Light Detection and Ranging (LIDAR) technology to map the city's surface down to the nearest centimeter. This enabled the city to identify areas most at risk from storm surges and developed detailed evacuation plans accordingly.

10.5. Community Engagement and Education

Equally important to these physical and technological changes are the social adaptations that cities are championing. The adoption of sustainable behaviors, informed by a deep understanding of climate change and its impacts, is indispensable in creating resilient cities.

Toronto's Live Green initiative provides a case in point. Through Live Green, the city has recruited a network of climate ambassadors who educate citizens about climate change, its impacts, and ways in which each individual can contribute to the city's resilience efforts.

In essence, resilience is not just about weathering the storm, but being able to thrive despite it. The strategies highlighted illustrate how cities are embracing innovation and adaptation measures with courage and determination. By doing so, these cities are not just preserving the well-being of the present day urbanites, but they are also ensuring the survival and prosperity of future generations to come. Resilience isn't just a city's response to climate change, but a testament to the indomitable spirit of humanity.

Chapter 11. Paving the Green Path: Policies Driving Urban Sustainability

Sustainability is not an over-night achievement, but a well-planned, strategic, and systematic drive orchestrated by multiple stakeholders. It commences with coherent policies aimed at restructuring existing frameworks and infusing innovative methods to replace them. In that regard, urban sustainability relies heavily on the efficacy and vision of various policies shaping it.

11.1. Institutional Framework and Role of Policies

Firstly, it is necessary to recognize the significance of institutional framework and the consequential role of policies. Sustainability policies at the urban level are often entrenched in larger national and international frameworks. These policies serve as the guiding pillar for a city's sustainability measures, addressing multiple facets - from waste management to energy conservation and pollution control. Together, they form an intricate latticework that sustainably drives urban development.

For instance, in Amsterdam, the city council's endeavors to enhance bike usage and eventually become car-free by 2030 has been directed by the country's overarching commitment to combat climate change. This conjugation of policies at different administrative levels enhances the effectiveness of the sustainability measures undertaken.

11.2. Green Transportation Policies

Green transportation policies are paramount when strategizing urban sustainability. The rise of car-centric cities has significantly amplified carbon emissions, leading to poor air quality and an increased carbon footprint. Green transportation policies aim to reduce dependability on automobiles, encouraging public transport and alternate methods of commuting.

Cities like Copenhagen have completely morphed their infrastructure to encourage cycling and walking, vowing to become the first carbon-neutral capital by 2025. Similarly, Bogota's TransMilenio Bus Rapid Transit (BRT) system has been a game-changer. In China, Shenzhen, known as the world's first city to introduce a fully electric bus fleet, demonstrates the power of proactive policies.

11.3. Policies for Energy-Efficient Buildings

According to the International Energy Agency (IEA), buildings and construction together account for almost 40% of global energy-related CO_2 emissions. Hence, policies fostering energy-efficient building designs and construction processes play a vital role. Several cities have enacted stringent regulations requiring new constructions to meet high energy efficiency standards.

For example, Vancouver's Zero Emissions Building Plan is a shining beacon. Launched in 2016, it mandates that by 2030, all new buildings will be carbon neutral. Singapore's Green Mark Scheme incentivizes green building design, scaling up energy efficiency to reduce the carbon footprint.

11.4. Waste Management and Circular Economy

Waste management is one of the most pressing issues urban habitats face today. Shifting towards a circular economy, through policies that encourage reusing, recycling, and composting can lead to waste reduction and sustainability.

San Francisco's Zero Waste Program sets a glorious precedent, aiming for zero waste by 2030. The policy enforces that all residents and businesses should properly sort their waste into recyclables, compostables, and trash to optimize waste management.

11.5. Climate Resilient Policies

Gradual urbanization has engendered distinct climate vulnerabilities. Climate resilience policies aim to ensure that cities are well-equipped to withstand and navigate climatic challenges. They encompass everything from resilient infrastructure development to disaster response strategies.

New York's OneNYC 2050 strategy involves building resiliency against severe weather events and rising sea levels. On the other end of the spectrum, Melbourne is concentrating on battling heat waves and possible drought scenarios under its strategy 'Climate Ready Melbourne'.

These concrete examples not only testify to the power of impactful policies but also serve as invaluable references for other cities. Indeed, the interwoven tapestry of these laws and regulations is what drives our cities towards the shared goal of sustainable existence. As various governments, stakeholders, and citizens engage collectively, the seed of urban sustainability, once sown, will inevitably sprawl into a vigorous green maze, brimming with life, resilience, and perpetual growth. As a result, the policies we create today will echo

as the triumphant proclamations of a greener, healthier, more
sustainable tomorrow.